CONTROLLER AREA NETWORK PROTOTYPING WITH ARDUINO

COPPERHILL TECHNOLOGIES

http://www.copperhilltech.com

CONTROLLER AREA NETWORK PROTOTYPING WITH ARDUINO
by Wilfried Voss

Published by
Copperhill Technologies Corporation
158 Log Plain Road
Greenfield, MA 01301
USA

ISBN-10: 1938581164
ISBN-13: 978-1-938581-16-8

http://www.copperhilltech.com

From the Author

It seems like a million years since I had a soldering iron in my hand and that I have been engaging in my most favorite programming activity, namely programming of embedded systems. In the past, I did shy away from the expenses that came with embedded programming, but with the emergence of inexpensive prototyping systems such as the Arduino or Raspberry Pi this concern doesn't exist anymore.

Add to this a virtually non-existing learning curve. With my Arduino Uno I ordered a book explaining Arduino Sketches, and I read it for about 30 minutes to scan through the most important information. Then it took maybe another 30 minutes to get my first application running.

I know, I am joining the enormous club of Arduino users who made and expressed the same experience, but that doesn't change the fact that the Arduino is the perfect environment for prototyping of embedded computer systems.

Naturally, with my knowledge of all kinds of Controller Area Network topics, I was eager to convert that knowledge into the real thing, namely a working CAN application. That CAN application will be the basics of an USB-to-CAN Gateway with CAN network monitoring and diagnostic features as explained in the chapters to follow. From here on, with the knowledge gained through this project, I encourage you to let your mind flow and extend the application. The possibilities are plenty. Enjoy!

However, before we get there let me explain the approach of writing this book: I could have engaged into writing many pages about Arduino basics, what it is, where it comes from, how to use it, etc., for the mere purpose of adding more pages and, as a result, being able to charge more money for my book. However, there are myriads of books on Arduino, Arduino Sketches, and Arduino Shields available in the market, and I won't waste your money or time. However, references to Arduino basics may appear but only in passing.

That being said, this book assumes some knowledge of the Arduino hardware and its programming.

It also assumes some basic knowledge of Controller Area Network (CAN). I will refer briefly to some aspects of CAN, but these are the mere basics of the actual protocol, just enough to understand the concept. In all truth, there is no need to understand all details of the protocol, since 100% of the protocol is implemented on a chip, the CAN controller. All we need to do in this book is to receive, transmit, and process data. The rest is up to your fantasy.

Nevertheless, the CAN protocol utilizes some ingenious features, and if you are interested in learning more, please refer to *A Comprehensible Guide to Controller Area Network* as mentioned in the literature appendix of this book.

Last, but not least, let me lose some words on my programming style that is definitely different than what you usually see.

I put great emphasis not only on readability of code; I also have debugging in mind when I write code. I am using a slightly modified version of the *Hungarian Notation*, meaning looking at a variable's or function's name provides you some information about its nature. For instance, the prefix *n* indicates an integer variable (e.g. nVariable). In addition, being familiar with a number of programming languages, I attempt to keep the best of all worlds. For instance, I add comments behind almost every bracket to indicate information such as *end if* or *end while*, etc., which helps identify program blocks. This may be helpful for *Visual Basic* programmers who are new to C/C++ programming.

Like under Visual Basic, my functions/routines start with either *Sub* (the return code is *void*) or *xFct* (where *x* indicates the type of the return code, for instance, *n* for integer).

About the Author

I am the author of the "Comprehensible Guide" series of technical literature covering topics like Controller Area Network (CAN), SAE J1939, Industrial Ethernet, and Servo Motor Sizing. I have worked in the CAN industry since 1997 and before that was a motion control engineer in the paper manufacturing industry. I have a master's degree in electrical engineering from the University of Wuppertal in Germany.

During the past years, I have conducted numerous seminars on industrial fieldbus systems such as CAN, CANopen, SAE J1939, Industrial Ethernet, and more during various *Real Time Embedded And Computing Conferences* (RTECC), ISA (Instrumentation, Systems, and Automation Society) conferences and various other events all over the United States and Canada.

I had the opportunity of traveling the world extensively, but settled in New England in 1989. I presently live in an old farmhouse in Greenfield, Massachusetts with my red-haired, green-eyed Irish-American wife and our son Patrick.

For more information on my works and to contact me, see my website at http://copperhilltech.com.

Contact the Author

Despite all efforts in preparing this book, there is always the possibility that some aspects or facts will not find everybody's approval, which prompts us, author and publisher, to ask for your feedback. If you would like to propose any amendments or corrections, please send us your comment. We look forward to any support in supplementing this book, and we welcome all discussions that contribute to making the topic of this book as thorough and objective as possible.

To submit amendments and corrections please log on to the author's website at http://copperhilltech.com/contact-us/ and leave a note.

Code And Projects Download

Any additional information created after the publishing date of this book plus project & source code (Arduino and Windows) are available as a free download through the author's website at http://copperhilltech.com/controller-area-network-can-prototyping-with-arduino/

Table of Content

1. Introduction to Controller Area Network

Controller Area Network (CAN) is a serial network technology that was originally designed for the automotive industry, especially for European cars, but has also become a popular bus in industrial automation as well as other applications. The CAN bus is primarily used in embedded systems, and as its name implies, is a network technology that provides fast communication among microcontrollers up to real-time requirements, eliminating the need for the much more expensive and complex technology of a Dual-Ported RAM.

CAN is a two-wire, half duplex, high-speed network system, that is far superior to conventional serial technologies such as RS232 in regards to functionality and reliability and yet CAN implementations are more cost effective.

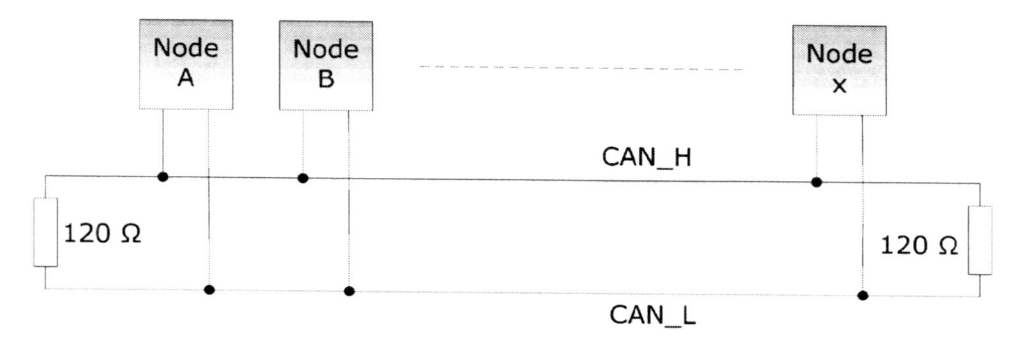

While, for instance, TCP/IP is designed for the transport of large data amounts, CAN is designed for real-time requirements and with its 1 MBit/sec baud rate can easily beat a 100 MBit/sec TCP/IP connection when it comes to short reaction times, timely error detection, quick error recovery and error repair.

CAN networks can be used as an embedded communication system for microcontrollers as well as an open communication system for intelligent devices. Some users, for example in the field of medical engineering, opted for CAN because they have to meet particularly stringent safety requirements.

Similar requirements had to be considered by manufacturers of other equipment with very high safety or reliability requirements (e.g. robots, lifts and transportation systems).

The greatest advantage of Controller Area Network lies in the reduced amount of wiring combined with an ingenious prevention of message collision (meaning no data will be lost during message transmission).

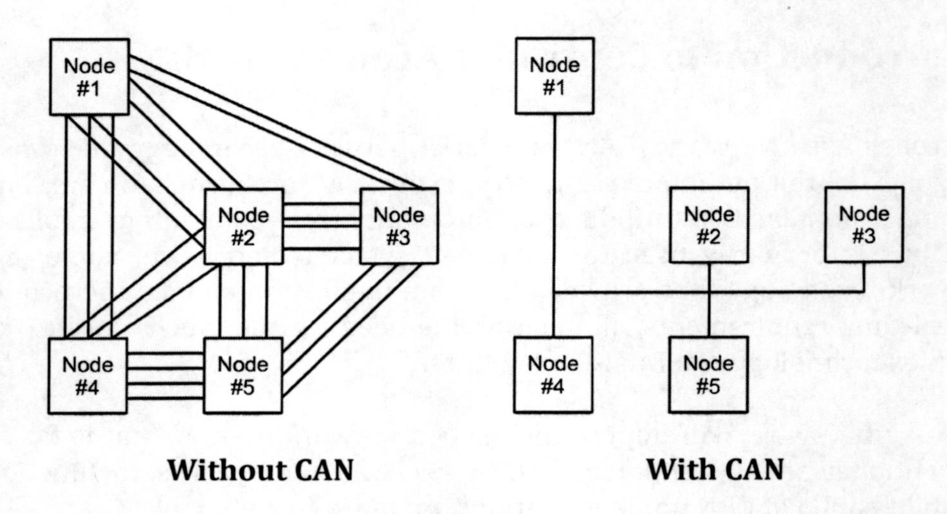

The following shows a need-to-know overview of CAN's technical characteristics.

Controller Area Network

- Is a serial networking technology for embedded solutions.
- Needs only two wires named CAN_H and CAN_L.
- Operates at data rates of up to 1 Megabit per second.
- Supports a maximum of 8 bytes per message frame.
- Does not support node IDs, only message IDs. One application can support multiple message IDs.
- Supports message priority, i.e. the lower the message ID the higher its priority.
- Supports two message ID lengths, 11-bit (standard) and 29-bit (extended).
- Does not experience message collisions (as they can occur under other serial technologies).
- Is not demanding in terms of cable requirements. Twisted-pair wiring is sufficient.

Note: *For more detailed information on CAN, please refer to "A Comprehensible Guide to Controller Area Network" as mentioned in the literature appendix of this book.*

2. Prototyping Hardware and its Variants

As I had mentioned earlier in this book, it is assumed that you have some basic knowledge of the Arduino, Arduino Sketches, and Arduino Shields. I will nevertheless take the opportunity of mentioning the prototyping hardware and its variants.

It is important to know that the Arduino, even though perfect for prototyping due to its low price and ease of programming, is not, in its bare form, an industrial-strength solution, not only in terms of environmental specs (e.g. temperature range, etc.) but also in terms of execution speed and memory resources.

Specifically, when it comes to CAN applications at 1 Mbit/sec and high data traffic, the Arduino may reach its limits quickly. There are, however, advanced and yet compatible alternatives to the Arduino as explained in the following chapters.

2.1 Arduino

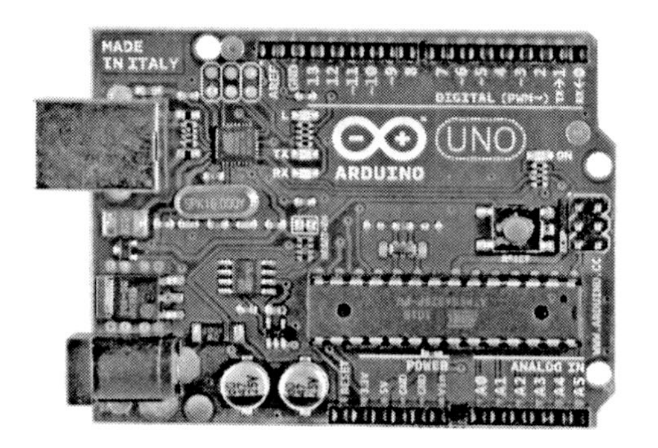

In order to develop and test the sample programs (sketches) as shown in this book, I used the Arduino Uno. The hardware consists of an open-source hardware board, usually designed around an 8-bit Atmel AVR microcontroller with 2 KB RAM (working memory), 32 KB Flash Memory (sketches) and 1 KB EEPROM (non-volatile).

These technical specifications are more than sufficient for basic prototyping of CAN applications and the proof of concept. However, to re-iterate the point, with growing demands for execution speed and extended functionality, the Arduino may quickly reach its limits.

Note: *All Arduino programs (sketches) as shown in this book were developed and tested with the Arduino Uno. There is no guarantee that these programs will work "as is" on any other compatible system.*

2.2 Intel Galileo

The Intel Galileo is a microcontroller board based on the Intel® Quark SoC X1000 Application Processor, a 32-bit Intel Pentium-class system on a chip. It is designed to be hardware and software pin-compatible with Arduino shields designed for the Uno R3.

The Galileo board is also software compatible with the Arduino software development environment, which should make usability and introduction a snap.

In addition to Arduino hardware and software compatibility, the Galileo board has several PC industry standard I/O ports and features to expand native usage and capabilities beyond the Arduino shield ecosystem. A full sized mini-PCI Express slot, 100Mb Ethernet port, Micro-SD slot, RS-232 serial port, USB Host port, USB Client port, and 8MByte NOR flash come standard on the board.

The CPU is a 400MHz 32-bit Intel® Pentium instruction set architecture (ISA)-compatible processor, and there is up to 8 MByte of Flash available. *(Source: Galileo Datasheet by Intel)*

For more information see: http://www.intel.com/content/www/us/en/do-it-yourself/galileo-maker-quark-board.html.

2.3 LeafLabs Maple Microcontroller Board

As similar as it may be to the Arduino, the differences are what really make the Maple stand out. It harnesses the power of a 32-bit ARM Cortex-M3 clocked at 72 MHz to push 39 GPIOs, 16 analog pins, 12-bit ADC resolution and 15 PWM pins at 16-bit resolution. In order to make sure you have plenty of programming room to flex that hardware, the Maple also provides 128k Flash and 20KB SRAM. All of this performance is delivered in the same form factor as the Arduino Pro.

If your current Arduino-based project is pushing against the performance limits of the ATmega, porting it over to Maple may be the fastest and easiest way to continue developing your project without starting from scratch.

By swapping the popular "avr-gcc" compiler with CodeSourcery's "arm-none-eabi-gcc," LeafLabs manages to provide a nearly identical programming experience to Arduino despite targeting a completely different architecture. Also, while some Arduino shields are incompatible due to certain capabilities being allocated to different pins, several of them are currently supported and there are more to come. There is also a guide available on the product page for porting Arduino libraries and source code over to Maple. *(Source: LeafLabs open electronis)*

For more information see: http://leaflabs.com/docs/hardware/maple.html

3. Arduino CAN Shields

Since Controller Area Network (CAN) is predominantly targeted at industrial solutions (versus the vastly more popular USB for non-industrial use such as home and lab), there aren't too many choices available in the market.

Through some research (i.e. browsing) I found two very similar solutions, and they both work with the same CAN library (as explained in a later chapter). Both solutions use the Microchip MCP2515 CAN controller. Also, both solutions are distributed through worldwide online resources.

3.1 Microchip MCP2515 CAN Controller

Microchip Technology's MCP2515 is a stand-alone Controller Area Network (CAN) controller that implements the CAN specification, version 2.0B. It is capable of transmitting and receiving both standard and extended data and remote frames. The MCP2515 has two acceptance masks and six acceptance filters that are used to filter out unwanted messages, thereby reducing the host MCUs overhead. The MCP2515 interfaces with microcontrollers (MCUs) via an industry standard Serial Peripheral Interface (SPI).

The features include two receive buffers with prioritized message storage, six 29-bit filters, two 29-bit masks, and three transmit buffers with prioritization and abort features. *(Source: Microchip Datasheet)*

Note: *CAN specification 2.0B refers to the capability of using standard CAN frames with 11-bit message identifier plus the extended format with a 29-bit message ID.*

To download the full MCP2515 datasheet log on to:
http://ww1.microchip.com/downloads/en/DeviceDoc/21801G.pdf

Both CAN shields as described in the following chapters utilize the Microchip MCP2551 CAN transceiver, which converts the internal TTL signals to a differential voltage as demanded by the CAN standard.

To download the full MCP2551 datasheet log on to:
http://ww1.microchip.com/downloads/en/DeviceDoc/21667f.pdf

3.2 Arduino CAN-Bus Shield by SK Pang electronics

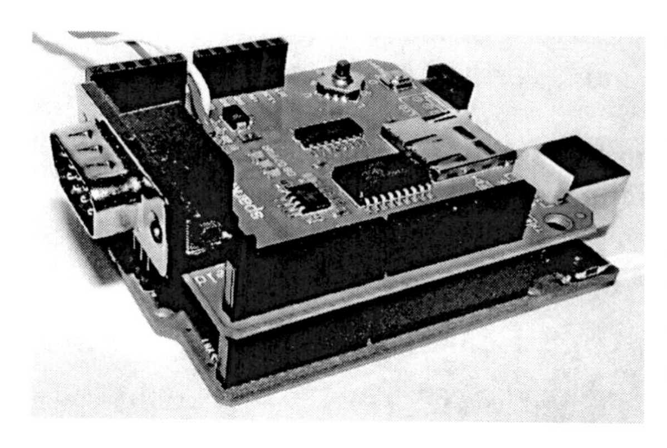

This shield by SK Pang electronics provides the Arduino CAN-Bus capability. As explained previously, it uses the Microchip MCP2515 CAN controller with MCP2551 CAN transceiver. The CAN connection is realized via a standard 9-way sub-D, however the pin assignment for CAN_H, CAN_L is not according to standard.

Note: *In all truth, there is no mandatory standard for pin assignment, but the industry uses pins 2 (CAN_L) and 7 (CAN_H) as a virtual standard.*

I recommend using the on-board CAN_L and CAN_H contacts to solder the CAN cable directly to the board.

The shield also comes with a uSD card holder, a serial LCD connector, and a connector for an EM406 GPS module, making this shield suitable for data logging application.

Features

- CAN v2.0B up to 1 Mb/s
- High speed SPI Interface (10 MHz)
- Standard and extened data and remote frames
- CAN connection via standard 9-pin sub-D connector
- As an option, power can be supplied to the Arduino by sub-D via resettable fuse and reverse polarity protection.
- Socket for EM406 GPS module
- Micro SD card holder
- Connector for serial LCD
- Reset button
- Joystick control menu navigation control
- Two LED indicator

Notes

- No cables included
- Header pins are not included; they must be ordered separately
- Pin assignment for CAN_H, CAN_L not according to standard

All technical information regarding the use of the CAN controller, uSD card holder, joystick, LEDs, etc. can be found on the company's wiki website at: https://code.google.com/p/skpang/

Ordering Information

To order the SK Pang ele3ctronics CAN shield, you can use the following resources (or browse for "Arduino CAN-BUS Shield" for further options):

Sparkfun - https://www.sparkfun.com/products/10039

SK Pang electronics - http://skpang.co.uk/catalog/arduino-canbus-shield-with-usd-card-holder-p-706.html

3.3 CAN-BUS Shield by Seeed Studio

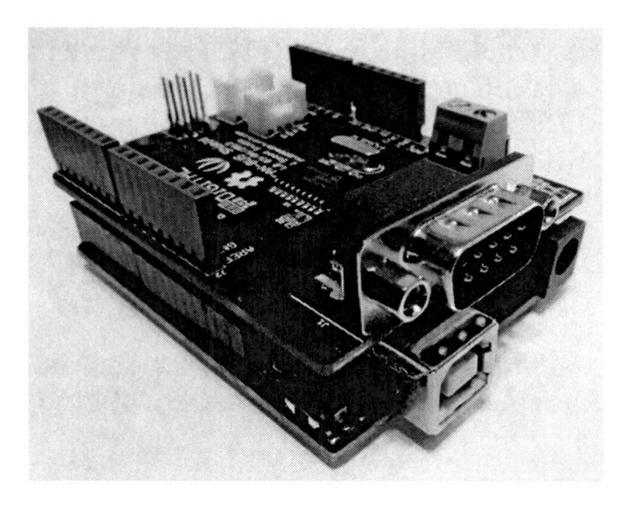

In terms of CAN capabilities, the shield by Seeek Studio provides the same functionality as the one by SK Pang electronics, however, it comes with a much lower price tag, because it does not have any additional components besides the CAN interface.

Over all, the device makes a solid impression, especially since the CAN connection is according to standard and in addition provides CAN connectivity through easily accessible terminals.

Features

- Implements CAN V2.0B at up to 1 Mb/s
- SPI Interface up to 10 MHz
- Standard (11 bit) and extended (29 bit) data and remote frames
- Two receive buffers with prioritized message storage
- Industrial standard 9 pin sub-D connector
- Two LED indicators

Notes

- No cables included

All technical information regarding the use of the CAN controller can be found on the company's wiki website at:
http://www.seeedstudio.com/wiki/CAN-BUS_Shield

Ordering Information

To order the Seeed Studio CAN shield, you can use the following resources (or browse for "Arduino CAN-BUS Shield" for further options):

Seeed Studio - http://www.seeedstudio.com/depot/CANBUS-shield-p-2256.html

Important to know: The Seeed Studio CAN bus shield has been undergoing some hardware changes to become compatible with systems such as the Arduino Mega 2560. The version 1.0 will work with the Arduino Uno, while all higher versions also work with the Mega 2560. This will also affect the code of the Arduino projects, specifically the line "MCP_CAN CAN0(10);" in the main module selecting the CS pin. That line must change to "MCP_CAN CAN0(9);" for all CAN bus shield versions above 1.0. I have added a comment in the corresponding section of the code.

4. Arduino CAN Sketches

The implementation of either one of the introduced CAN-BUS Shields and the corresponding CAN sketches went surprisingly smooth when paired with the right library software.

I found several source codes for accessing the MCP2515 CAN controller, but most of them didn't even pass the initial quality control phase (I read the code first before I use it). One of the quality criteria was the support for 29-bit CAN message identifiers (CAN 2.0B Compatibility), which is mandatory when it comes to implementing, for instance, the SAE J1939 vehicle network protocol. Some software samples I found were just literally "samples" and they left ample room for guessing games.

I was most pleased by the MCP2515 Library by Cory Fowler, which can be found at https://github.com/coryjfowler/MCP2515_lib

This library is compatible with any shield or CAN interface that uses the MCP2515 CAN protocol controller.

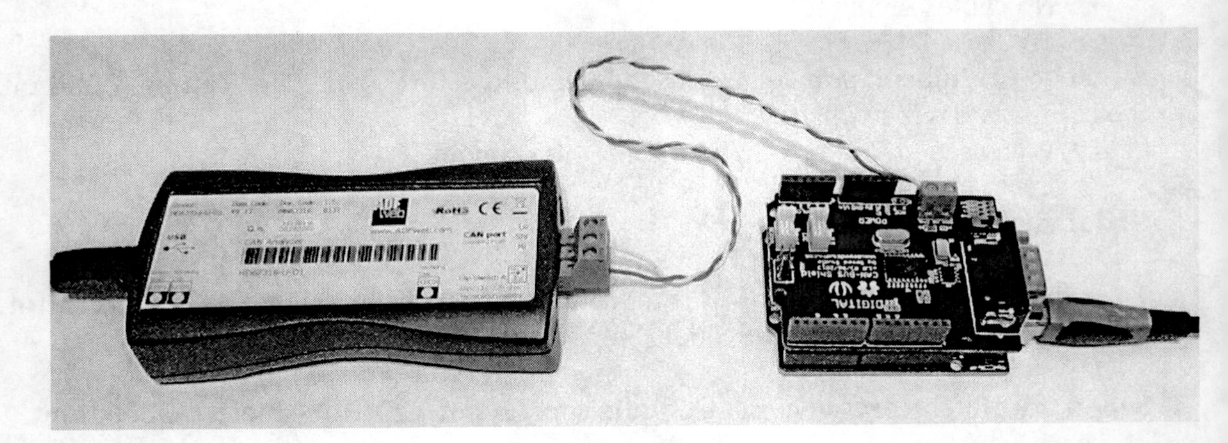

In order to test and verify the proper transmission and reception of CAN messages, I used the ADFweb CAN-to-USB gateway with its Windows interface.

Note: *In order to test a CAN application, you need at least two CAN nodes to establish a network communication. The second node can be another Arduino with CAN shield or (if the budget allows) another CAN device with CAN data monitoring capabilities.*

4.1 The MCP2515 Library

As with any serial networking controller, the essential functions are:

1. Initialization
2. Read Data
3. Write Data
4. Check Status

In case of the MCP2515 library, these functions are represented by:

1. Initialization: CAN0.begin
2. Read Data: CAN0.readMsgBuf
   incl. CAN0.checkReceive, CAN0.getCanId
3. Write Data: CAN0.sendMsgBuf
4. Check Status: CAN0.checkError

4.1.1 Function Calls

Function:	**CAN0.begin**
Purpose:	Initializes the CAN controller and sets the speed (baud rate)
Parameter:	CAN_5KPS ... CAN_1000KPS (See mcp_can_dfs.h)
Return Code:	CAN_OK = Initialization okay CAN_FAILINIT = Initialization failed

Function:	**CAN0.checkReceive**
Purpose:	Check if message was received
Parameter:	None

Return Code: CAN_MSGAVAIL = Message available
CAN_NOMSG = No message

Function: **CAN0.readMsgBuf**
Purpose: Read the message buffer
Parameter: nMsgLen returns the message length (number of data bytes)
nMsgBuffer returns the actual message
Return Code: None

Function: **CAN0.getCANId**
Purpose: Retrieves the ID of the received message
Parameter: None
Return Code: m_nID = Message ID

Function: **CAN0. sendMsgBuf**
Purpose: Send a message buffer
Parameter: id = Message ID
ext = CAN_STDID (11-bit ID) or CAN_EXTID (29-bit ID)
len = Number of data bytes (0...8)
buf = Message buffer
Return Code: None

Function: **CAN0.checkError**
Purpose: Checks CAN controller for errors
Parameter: None
Return Code: CAN_OK = Status okay
CAN_CTRLERROR = Error

There are further functions, among others, for message filtering and settings masks, and they are worth being checked out for more sophisticated functions, but they are not necessary for simple CAN communication tasks.

4.1.2 Implementation

The implementation of the MPC2515 library is fairly easy: Open Arduino, create a new file, then use the menu items *Sketch->Add File...* to include the following files to the project:

- mcp_can.cpp
- mcp_can.h
- mcp_can_dfs.h

In the Arduino project file add the following on top:

```
#include "mcp_can.h"
#include <SPI.h>
MCP_CAN CAN0(10);
```

Let me repeat here: The Seeed Studio CAN bus shield has been undergoing some hardware changes to become compatible with systems such as the Arduino Mega 2560. The version 1.0 will work with the Arduino Uno, while all higher versions also work with the Mega 2560. This will also affect the code of the Arduino projects, specifically the line “MCP_CAN CAN0(10);” in the main module selecting the CS pin. That line must change to “MCP_CAN CAN0(9);” for all CAN bus shield versions above 1.0.

You are now ready to go. The following chapters will describe how to implement the function calls.

4.2 CAN Programming

The most exciting part about this project is when it comes to the point where two CAN nodes communicate with each other. I started off with writing a simple program that sent messages that were received by my USB-to-CAN gateway and its Windows monitoring software. From there on, I extended the program to also receive CAN messages and display them on the Arduino serial monitor.

In a later chapter, I will also show a Windows programming example that establishes a communication with the Arduino.

4.2.1 Simple CAN Shield Test

The following represents a very simple CAN test program that periodically (i.e. every 1 second) sends out a CAN message with a 29-bit identifier at a baud rate of 250 kbit/sec.

```
// Simple CAN Shield Test
#include "mcp_can.h"
#include <SPI.h>
MCP_CAN CAN0(10); // Set CS to pin 10

unsigned char stmp[8] = {0x30, 0x31, 0x32, 0x33, 0x34, 0x35, 0x36, 0x37};

//  SYSTEM: Setup routine runs on power-up or reset
void setup() {

  // Set the serial interface baud rate
```

```
  Serial.begin(9600);

  // Initialize the CAN controller
  // Baud rates are defined in mcp_can_dfs.h
  if (CAN0.begin(CAN_250KBPS) == CAN_OK)
    Serial.print("CAN Init OK.\n\r\n\r");
  else
    Serial.print("CAN Init Failed.\n\r");

}// end setup

// Main Loop - Arduino Entry Point
void loop()
{
  // Send data:  id = 0x1FF, extended frame, data len = 8, stmp: data buf
  // ID mode (11/29 bit) defined in mcp_can_dfs.h
  CAN0.sendMsgBuf(0x1FF, CAN_EXTID, 8, stmp);

  // Run in 1 sec interval
  delay(1000);

}// end loop
```

While the code is short and sufficiently self-explanatory, let me explain the steps taken in the program.

In the ***setup()*** routine, the program initializes the serial interface (USB) to a baud rate of 9600 bps (Please make sure that your Arduino serial monitor is set to the same rate).

It then initializes the CAN controller to a data transmission rate of 250 kbits/sec and displays possible error messages on the Arduino serial monitor.

In the main ***loop()*** routine, the program sends an 8-byte CAN message using an ID of 0x1FF in extended messaging format (29-bit message ID). The actual message (unsigned char stmp[8] in this example) contains a string from ASCII 0 to 7, which can be easily spotted when using a data monitoring software.

Using my test conditions, the message was received through the USB-to-CAN gateway and displayed under Windows:

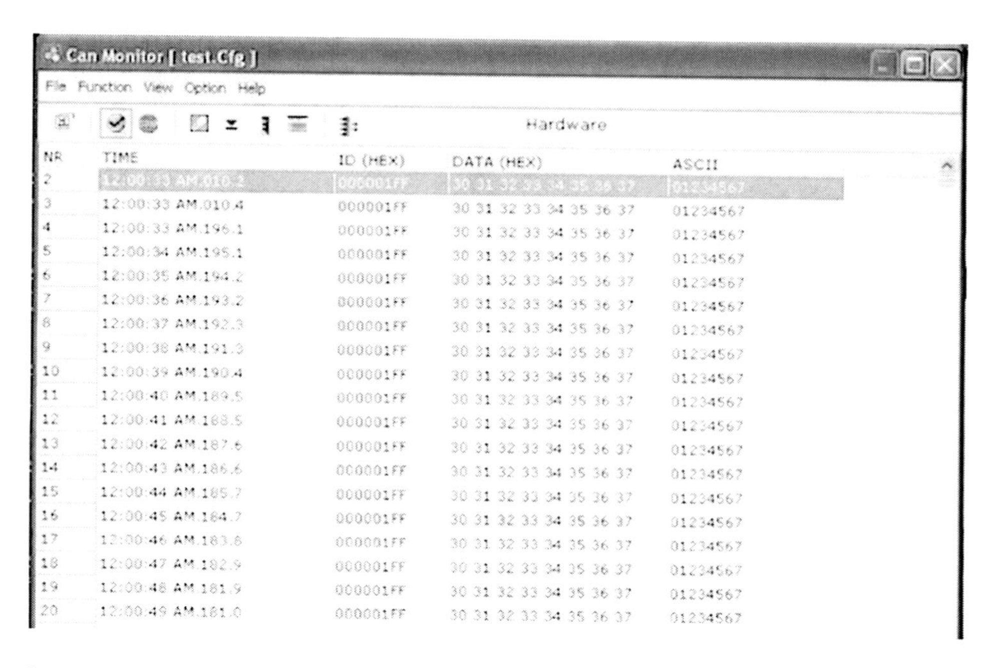

NR	TIME	ID (HEX)	DATA (HEX)	ASCII
2	12:00:33 AM.010.2	000001FF	30 31 32 33 34 35 36 37	01234567
3	12:00:33 AM.010.4	000001FF	30 31 32 33 34 35 36 37	01234567
4	12:00:33 AM.196.1	000001FF	30 31 32 33 34 35 36 37	01234567
5	12:00:34 AM.195.1	000001FF	30 31 32 33 34 35 36 37	01234567
6	12:00:35 AM.194.2	000001FF	30 31 32 33 34 35 36 37	01234567
7	12:00:36 AM.193.2	000001FF	30 31 32 33 34 35 36 37	01234567
8	12:00:37 AM.192.3	000001FF	30 31 32 33 34 35 36 37	01234567
9	12:00:38 AM.191.3	000001FF	30 31 32 33 34 35 36 37	01234567
10	12:00:39 AM.190.4	000001FF	30 31 32 33 34 35 36 37	01234567
11	12:00:40 AM.189.5	000001FF	30 31 32 33 34 35 36 37	01234567
12	12:00:41 AM.188.5	000001FF	30 31 32 33 34 35 36 37	01234567
13	12:00:42 AM.187.6	000001FF	30 31 32 33 34 35 36 37	01234567
14	12:00:43 AM.186.6	000001FF	30 31 32 33 34 35 36 37	01234567
15	12:00:44 AM.185.7	000001FF	30 31 32 33 34 35 36 37	01234567
16	12:00:45 AM.184.7	000001FF	30 31 32 33 34 35 36 37	01234567
17	12:00:46 AM.183.8	000001FF	30 31 32 33 34 35 36 37	01234567
18	12:00:47 AM.182.9	000001FF	30 31 32 33 34 35 36 37	01234567
19	12:00:48 AM.181.9	000001FF	30 31 32 33 34 35 36 37	01234567
20	12:00:49 AM.181.0	000001FF	30 31 32 33 34 35 36 37	01234567

While this program may not be very useful without a CAN monitoring software (meaning you can't see the result), in the least it demonstrates how simple CAN programming can be.

4.2.2 Extended CAN Shield Test

In this next, extended example, we use the same program as shown in the previous chapter but add a CAN receiving routine to it. The result, i.e. the received messages, will be displayed through the Arduino serial monitor.

```
// Simple CAN Shield Test
#include <stdlib.h>
#include "mcp_can.h"
#include <SPI.h>
MCP_CAN CAN0(10);                                        // Set CS to pin 10

// Test message
unsigned char stmp[8] = {0x30, 0x31, 0x32, 0x33, 0x34, 0x35, 0x36, 0x37};

//  SYSTEM: Setup routine runs on power-up or reset
void setup() {

  // Set the serial interface baud rate
  Serial.begin(9600);

  // Initialize the CAN controller
  // Baud rates defined in mcp_can_dfs.h
  if (CAN0.begin(CAN_250KBPS) == CAN_OK)
    Serial.print("CAN Init OK.\n\r\n\r");
  else
```

```
    Serial.print("CAN Init Failed.\n\r");

}// end setup

// Main Loop - Arduino Entry Point
void loop()
{
  // Declarations
  byte nMsgLen = 0;
  byte nMsgBuffer[8];
  char sString[4];

  // Send out a test message
  // Send data:  id = 0x1FF, extended frame, data len = 8, stmp: data buf
  // ID mode (11/29 bit) defined in mcp_can_dfs.h
  CAN0.sendMsgBuf(0x1FF, CAN_EXTID, 8, stmp);

  // Check for a message
  if(CAN0.checkReceive() == CAN_MSGAVAIL)
  {
    // Read the message buffer
    CAN0.readMsgBuf(&nMsgLen, &nMsgBuffer[0]);
    INT32U nMsgID = CAN0.getCanId();

    // Print message ID to serial monitor
    Serial.print("Message ID: 0x");
    if(nMsgID < 16) Serial.print("0");
    Serial.print(itoa(nMsgID, sString, 16));
    Serial.print("\n\r");

    // Print data to serial monitor
    Serial.print("Data: ");
    for(int nIndex = 0; nIndex < nMsgLen; nIndex++)
    {
      Serial.print("0x");
      if(nMsgBuffer[nIndex] < 16) Serial.print("0");
      Serial.print(itoa(nMsgBuffer[nIndex], sString, 16));
      Serial.print(" ");
    }// end for
    Serial.print("\n\r\n\r");

  }// end if

  // Run in 1 sec interval
  delay(1000);

}// end loop
```

Obviously, the program has grown compared to the previous one, but most of the added code is used for the data display on the Arduino serial monitor.

First, note on top the line ***#include <stdlib.h>.*** The stdlib.h file allows us to convert integer data into ASCII, which is necessary for the data display.

The ***setup()*** routine remains the same as it was in the previous example.

In the ***loop()*** routine, we first declare some variables for the message reception and code conversion. We still send out the same message as before by calling the ***CAN0.sendMsgBuf()*** function.

Next, we check for the reception of a CAN message, and if that is the case, we read the message into the assigned buffer and retrieve the message ID. The following code is all about converting the received data into a human-readable format (ASCII) and display it on the Arduino serial monitor.

Last, but not least, we halt the system for one second (1000 milliseconds). Naturally, under real-life conditions, this delay is not reasonable, since there can occur literally thousands of messages per one second. However, this code is meant merely as a demo sample that proves that the actual CAN communication can be accomplished with very little code.

If you load this program onto two separate Arduinos with CAN shield, you have not only accomplished a full CAN network, you can also see the CAN messages as they are exchanged between the two nodes.

Note: *It may sound obvious, but please make sure, in case you use more than one CAN node, that all nodes are initialized with the same baud rate. Using different baud rates is the most common cause when data communication fails.*

4.2.3 A Simple CAN Network Monitoring and Diagnostics Program

The Arduino board in combination with the CAN shield provides the hardware for a full-fledged CAN network monitoring tool, and this next Arduino program is a first step in that direction.

However, before we get into more detail, let me issue some warnings regarding possible restrictions of the system:

- The MPC2515 has only two receive buffers, which limits the system's capabilities to respond in a timely fashion while receiving and processing the data traffic. For high-speed, high-busload applications, it is recommended to use the message filter functions to reduce the processing load on the CPU.

- Besides the limited processing speed of the 8-bit CPU, the Arduino comes with only 32 kByte program memory, which is sufficient for a great number of small applications. However, when it comes to more demanding tasks such as a CAN monitoring tool, the memory resources may be exhausted quicker than expected. For instance, the following application already uses roughly 20 percent of the total memory space, and it provides only a very rudimentary version of a monitoring tool.

In all consequence, if you are serious about creating a more-or-less professional application, you might want to consider alternative hardware solutions as discussed in a previous chapter.

In order to create a CAN monitoring system, we need to:

1. Receive CAN messages and display them
2. Be able to enter CAN messages and transmit them

With the previous two programming samples in mind, we have already accomplished step #1, but the next step (entering CAN messages) needs a bit more work.

The idea is to enter the CAN message into Arduino's serial monitor and transmit the result by clicking the *Send* button. In order to accomplish that, we need to follow a data entry format as shown in the following.

Command: **Send CAN Message (11 bit)**
Description: Node receives a message and transmits it into the CAN bus
Format: #SM id n dd dd....

id = Message ID (2 bytes, hex)
n = Number of bytes (1 byte, 0 to 8)
d = data bytes (hex, up to eight bytes)

Example:
#SM 01FF 8 30 31 32 33 34 35 36 37

In this previous example, we design a CAN message with an ID of 01FF and a data length of 8 bytes. These 8 bytes are represented by the number 30 (hex) to 37, which is the equivalent of ASCII-0 to ASCII-7.

While the basic functionality of sending and receiving CAN messages remains the same, the program size and complexity has, naturally, grown. Most of the code, however, is being used for conversion between hex and ASCII formats (for readability) and some rudimentary syntax check.

Note: *The data entry in this following programming sample is not fool-proof, meaning, while the program does some syntax checks, it is still possible that incorrect data entries will still be interpreted as valid CAN message formats.*

Also, this example still uses 9600 baud for the communication with Arduino's serial monitor. A faster transmission speed is recommended for CAN networks with high data traffic.

```
// Simple CAN Shield Test
#include <stdlib.h>
#include "mcp_can.h"
#include <SPI.h>
MCP_CAN CAN0(10);                                          // Set CS to pin 10

// Constants
#define MAX_CMD_LENGTH 60

#define CR "\n\r"
#define CRCR "\n\r\n\r"

//  SYSTEM: Setup routine runs on power-up or reset
void setup() {

  // Set the serial interface baud rate
  Serial.begin(9600);

  // Initialize the CAN controller
  // Baud rates defined in mcp_can_dfs.h
  if (CAN0.begin(CAN_250KBPS) == CAN_OK)
    Serial.print("CAN Init OK.\n\r\n\r");
  else
    Serial.print("CAN Init Failed.\n\r");

}// end setup

// Main Loop - Arduino Entry Point
void loop()
{
  // Check for a received CAN message and print it to the Serial Monitor
  SubCheckCANMessage();

  // Check for a command from the Serial Monitor and send message as entered
  SubSerialMonitorCommand();

}// end loop

// -----------------------------------------------------------------------
```

```
// Check for CAN message and print it to the Serial Monitor
// ---------------------------------------------------------------------------
void SubCheckCANMessage(void)
{
  // Declarations
  byte nMsgLen = 0;
  byte nMsgBuffer[8];
  char sString[4];

  if(CAN0.checkReceive() == CAN_MSGAVAIL)
  {
    // Read the message buffer
    CAN0.readMsgBuf(&nMsgLen, &nMsgBuffer[0]);
    INT32U nMsgID = CAN0.getCanId();

    // Print message ID to serial monitor
    Serial.print("Message ID: 0x");
    if(nMsgID < 16) Serial.print("0");
    Serial.print(itoa(nMsgID, sString, 16));
    Serial.print("\n\r");

    // Print data to serial monitor
    Serial.print("Data: ");
    for(int nIndex = 0; nIndex < nMsgLen; nIndex++)
    {
      Serial.print("0x");
      if(nMsgBuffer[nIndex] < 16) Serial.print("0");
      Serial.print(itoa(nMsgBuffer[nIndex], sString, 16));
      Serial.print(" ");
    }// end for
    Serial.print(CRCR);

  }// end if

}// end subCheckCANMessage

// ---------------------------------------------------------------------------
// Check for command from Serial Monitor
// ---------------------------------------------------------------------------
void SubSerialMonitorCommand()
{
  // Declarations
  char sString[MAX_CMD_LENGTH+1];
  bool bError = true;

  unsigned long nMsgID = 0xFFFF;
  byte nMsgLen = 0;
  byte nMsgBuffer[8];

  // Check for command from Serial Monitor
  int nLen = nFctReadSerialMonitorString(sString);

  if(nLen > 0)
  {
    // A string was received from serial monitor
    if(strncmp(sString, "#SM ", 4) == 0)
    {
      // The first 4 characters are acceptable
```

```
      // We need at least 10 characters to read the ID and data number
      if(strlen(sString) >= 10)
      {
        // Determine message ID and number of data bytes
        nMsgID = lFctCStringLong(&sString[4], 4);
        nMsgLen = (byte)nFctCStringInt(&sString[9], 1);

        if(nMsgLen >=0 && nMsgLen <=8)
        {
           // Check if there are enough data entries
           int nStrLen = 10 + nMsgLen * 3;  // Expected msg length
           if(strlen(sString) >= nStrLen) // Larger length is acceptable
           {
             int nPointer;
             for(int nIndex = 0; nIndex < nMsgLen; nIndex++)
             {
               nPointer = nIndex * 3;  // Blank character plus two numbers
               nMsgBuffer[nIndex] =
                           (byte)nFctCStringInt(&sString[nPointer + 11], 2);
             }// end for

             // Reset the error flag
             bError = false;

             // Everything okay; send the message
             CAN0.sendMsgBuf(nMsgID, CAN_STDID, nMsgLen, nMsgBuffer);

             // Repeat the entry on the serial monitor
            Serial.print(sString);
            Serial.print(CRCR);

           }// end if

       }// end if

      }// end if

    }// end if

    // Check for entry error
    if(bError == true)
    {
      Serial.print("???: ");
      Serial.print(sString);
      Serial.print(CR);
    }

  }// end if

}// end SubSerialMonitorCommand

// ------------------------------------------------------------------------
// Read message from Serial Monitor
// ------------------------------------------------------------------------
// Returns string length
//
byte nFctReadSerialMonitorString(char* sString)
{
```

```
  // Declarations
  byte nCount;

  nCount = 0;

  if(Serial.available() > 0)
  {
    Serial.setTimeout(100);
    nCount = Serial.readBytes(sString, MAX_CMD_LENGTH);
  }// end if

  // Terminate the string
  sString[nCount] = 0;

  return nCount;

}// end nFctReadSerialMonitorString

// ------------------------------------------------------------------------
// Convert string into int
// ------------------------------------------------------------------------
// Note: nLen MUST be between 1 and 4
//
// Returns integer value (-1 indicates an error in the string)
//
int nFctCStringInt(char *sString, int nLen)
{
  // Declarations
  int nNum;
  int nRetCode = 0;

  // Check the string length
  if(strlen(sString) < nLen)
    nRetCode = -1;
  else
  {
    // String length okay; convert number
    int nShift = 0;
    for(int nIndex = nLen - 1; nIndex >=0; nIndex--)
    {
      if(sString[nIndex] >= '0' && sString[nIndex] <= '9')
        nNum = int(sString[nIndex] - '0');
      else if(sString[nIndex] >= 'A' && sString[nIndex] <= 'F')
        nNum = int(sString[nIndex] - 'A') + 10;
      else goto nFctCStringInt_Ret;

      nNum = nNum << (nShift++ * 4);
      nRetCode = nRetCode + nNum;

    }// end for

  }// end else

  // Return the result
nFctCStringInt_Ret:

  return nRetCode;
```

```
}// end nFctCStringInt

// ------------------------------------------------------------------------
// Convert string into unsigned long
// ------------------------------------------------------------------------
// Note: nLen MUST be between 1 and 4
//
// Returns integer value (-1 indicates an error in the string)
//
unsigned long lFctCStringLong(char *sString, int nLen)
{
  // Declarations
  unsigned long nNum;
  unsigned long nRetCode = 0;

  // Check the string length
  if(strlen(sString) < nLen)
    nRetCode = -1;
  else
  {
    // String length okay; convert number
    unsigned long nShift = 0;
    for(int nIndex = nLen - 1; nIndex >=0; nIndex--)
    {
      if(sString[nIndex] >= '0' && sString[nIndex] <= '9')
        nNum = int(sString[nIndex] - '0');
      else if(sString[nIndex] >= 'A' && sString[nIndex] <= 'F')
        nNum = int(sString[nIndex] - 'A') + 10;
      else goto lFctCStringLong_Ret;

      nNum = nNum << (nShift++ * 4);
      nRetCode = nRetCode + nNum;

    }// end for

  }// end else

  // Return the result
lFctCStringLong_Ret:

  return nRetCode;

}// end lFctCStringLong
```

Note: *This programming example, unlike the first two samples in this book, is based on the use of an 11-bit message identifier, paying tribute to the majority of CAN applications.*

Without going into the last detail, here is a brief description of the code:

The ***setup()*** function remains the same as in the first two programming examples in this book, i.e. it handles the initialization of the serial connection and the CAN controller.

The ***loop()*** routine, however, looks extremely simple, but that only means that the major part of the functionality has been distributed to a number of new functions.

Inside ***loop()*** are only two function calls:

1. ***SubCheckCANMessage()*** checks for a received CAN message and displays it on the Arduino serial monitor.

2. ***SubSerialMonitorCommand()*** receives a string from Arduino's serial monitor, achieves some rudimentary syntax check, and sends out the CAN message.

The remaining function calls are:

- ***nFctReadSerialMonitorString()*** reads the data format string as entered by the user and returns the string length.

- ***nFctCStringInt()*** converts a string into integer and returns the integer data.

- ***lFctCStringLong()*** converts a string into long and returns the long data.

Note: *Unlike C#, the C and C++ programming languages provide only limited support for data conversion, and sometimes writing your own conversion functions fits your application needs better than the provided library functions.*

The following shows screen shots taken trough a session with this programming example:

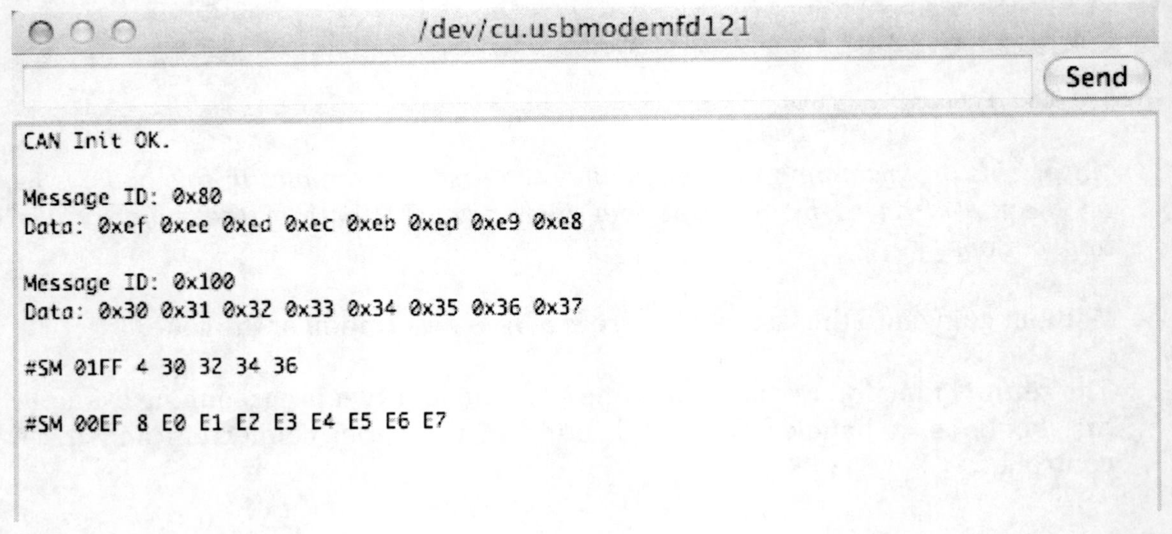

First, we received two CAN messages (IDs 0x80 and 0x100), then we sent two CAN messages (IDs 01FF and 00EF).

For this operation, I used my standard test configuration (i.e. USB-to-CAN gateway with Windows monitoring tool as the second CAN node), but from here on, it is possible to use two Arduinos with CAN shield running the same application.

In order to extend the functionality of this programming example, the following commands would be helpful to provide a full-fledged monitoring and diagnostics tool:

- **CAN Start/Stop** – Starts or stops displaying messages on the Arduino serial monitor.
- **CAN Baud Rate** – Modify the CAN baud rate.
- **Request CAN Settings** – Reports the current settings such as baud rate and message ID mode.
- **Send CAN Message** in 29-bit format.
- **Add CAN Message Filter**
- **Delete CAN Message Filter**
- **Delete All CAN Message Filters**

And yes, there are multiple possibilities of extending this program toward a really professional version. However, what the Arduino cannot provide is a professionally looking graphical user interface, and this is where the existing USB connection to a PC opens the door to more possibilities.

4.3 CAN Network Monitoring under Windows

While programming the Arduino can be exciting (especially since everything works so smoothly), the real fun comes when you can extend the Arduino's reach to a PC running Windows.

<u>Note</u>: *My apologies to all Mac and LINUX users for bringing a Windows programming example, but there is no better programming than using C# under Microsoft's Visual Studio. I have enjoyed programming under OS-X and LINUX, but when it comes to producing quick and effective programming examples, I prefer to stay with Visual Studio. However, the experienced programmer should be able to replicate the functionality of the serial monitor.*

To learn more about serial port programming (RS-232 and USB) under LINUX see http://www.teuniz.net/RS-232/. I consider this by far the most professional application for serial ports under LINUX. It also suits Windows applications but is primarily meant for compilers inferior to Visual Studio and/or for programming embedded systems.

In the following we assume that you have the Arduino USB driver installed under your Windows machine. The driver is automatically installed with the Arduino development environment.

As I have mentioned in my note, I am using Microsoft's Visual Studio 2012, and I have designed the following GUI that may look very familiar to the Arduino developer. Basically, this very simple program is a replica of the Arduino serial monitor.

The screen elements are a textbox for data entry, a command button to send the entry to the Arduino, and another larger text box to display the data coming from the Arduino. Last, but not least, there is a combobox displaying all available USB COM ports (It is your task to determine the proper USB port; there is no auto detection).

What the program does not provide is the baud rate settings, which has been hard-coded as 9600 baud into the program but can be modified easily. Of course, this is not the professional way of doing it, but, after all, this programming sample serves as an example on reading/sending messages from/to the Arduino. In regards to the envisioned extended CAN network monitoring and diagnostics tool, you will need more and different screen elements, and the baud rate settings should be part of that project.

All screen elements in this project stick with their default settings, however with a few exceptions as shown in the following:

Element	**Name**	**Modified Property**	**Events**
Form	Form1	Text = "Serial Monitor"	-
Text	txtSend	-	-
Text	txtReceived	Multiline = True Scrollbars = Vertical	
Button	btnSend	Text="Send"	Click
ComboBox	cboCOMPort	-	SelectedIndexChanged

The following shows the C# program listing (the entire program is within the form):

```
using System;
using System.Collections.Generic;
using System.ComponentModel;
using System.Data;
using System.Drawing;
using System.Linq;
using System.Text;
using System.Threading.Tasks;
using System.Windows.Forms;
using System.IO;
using System.IO.Ports;
using System.Threading;

namespace USBAccess
{
    public partial class Form1 : Form
    {
        // Constants
        public const int REC_BUFFER_SIZE = 500;
        public const int READ_TIMEOUT = 500;
        public const int WRITE_TIMEOUT = 500;
        public const int REC_BUFFER_FILLTIME = 80;

        public static SerialPort _serialport;

        public Form1()
        {
            InitializeComponent();
```

```
    string[] sPorts = new string[20];
    sPorts = SerialPort.GetPortNames();

    for (int nIndex = 0; nIndex < sPorts.Length; nIndex++)
        cboCOMPort.Items.Add(sPorts[nIndex]);

}// end Form1

// SetTextDeleg
// ------------
private delegate void SetTextDeleg(string text);

// sp_DataReceived
// ---------------
void sp_DataReceived(object sender, SerialDataReceivedEventArgs e)
{
    // Set the receive buffer size
    char[] sRecData = new char[REC_BUFFER_SIZE + 1];


    // Give the hardware some time to receive the whole message
    Thread.Sleep(REC_BUFFER_FILLTIME);

    try
    {
        int nBytes = _serialport.BytesToRead;

        // Read the string
        int nIndex;

        for (nIndex = 0; nIndex < nBytes; nIndex++)
        {
            int nRec = _serialport.ReadByte();
            sRecData[nIndex] = (char)nRec;

        }// end for

        sRecData[nIndex] = (char)0; // Terminate the string
        string sStr = new string(sRecData);

        // In case of RS232, this line causes a timeout,
        // meaning no data is being received
        this.BeginInvoke(new SetTextDeleg(si_DataReceived),
                                        new object[] { sStr });
    }
    catch (TimeoutException) { }

}// end _serialport_DataReceived

// si_DataReceived
// ---------------
private void si_DataReceived(string data)
{
    if(txtReceived.TextLength == 0)
        txtReceived.Text = data;
    else
        txtReceived.Text += "\n\r" + data;
```

```
            // Set cursor to end of screen
            txtReceived.SelectionStart = txtReceived.TextLength;
            txtReceived.ScrollToCaret();
            txtReceived.Refresh();

        }// end si_DataReceived

        // btnSend_Click
        // --------------
        private void btnSend_Click(object sender, EventArgs e)
        {
            // Make sure the serial port is open before trying to write
            try
            {
                if (!(_serialport.IsOpen))
                    _serialport.Open();

                if (txtSend.Text.Length > 0)
                    _serialport.Write(txtSend.Text);
                else
                    MessageBox.Show("Please enter a message to be sent.",
                                                        "Attention!");
            }
            catch (Exception ex)
            {
                MessageBox.Show("Error opening/writing to serial port." +
                                                ex.Message, "Error!");
            }

        }// end btnSend_Click

        // Event : cboCOMPort_SelectedIndexChanged
        //----------------------------------------
        private void cboCOMPort_SelectedIndexChanged(object sender,
                                                        EventArgs e)
        {
            // Define the serial port for the USB device
            _serialport = new SerialPort(cboCOMPort.SelectedItem.ToString(),
                                    9600, Parity.None, 8, StopBits.One);
            _serialport.Handshake = Handshake.None;

            // Set the read/write timeouts
            _serialport.ReadTimeout = READ_TIMEOUT;
            _serialport.WriteTimeout = WRITE_TIMEOUT;
            _serialport.ReadBufferSize = REC_BUFFER_SIZE;
            _serialport.Open();

            _serialport.DataReceived +=
                        new SerialDataReceivedEventHandler(sp_DataReceived);

        }// end cboCOMPort_SelectedIndexChanged

    }// end class

}// end namespace
```

Reference: *The handling of the USB port is based on an article by Ryan Alford (with added content by Arjun Walmiki, Gregory Krzywoszyja and Mahesh Chand) at: http://www.c-sharpcorner.com/uploadfile/eclipsed4utoo/communicating-with-serial-port-in-C-Sharp/*

At program start, the user first needs to select the applicable USB COM port, which initializes the port (SelectedIndexChanged event).

Beyond that, the program functions as a simple USB terminal: Messages are typed in the top text box and sent by clicking on the *Send* command button. The larger text box displays the received data.

The following shows screen shots taken through a session with our Arduino CAN Network Monitoring and Diagnostics program:

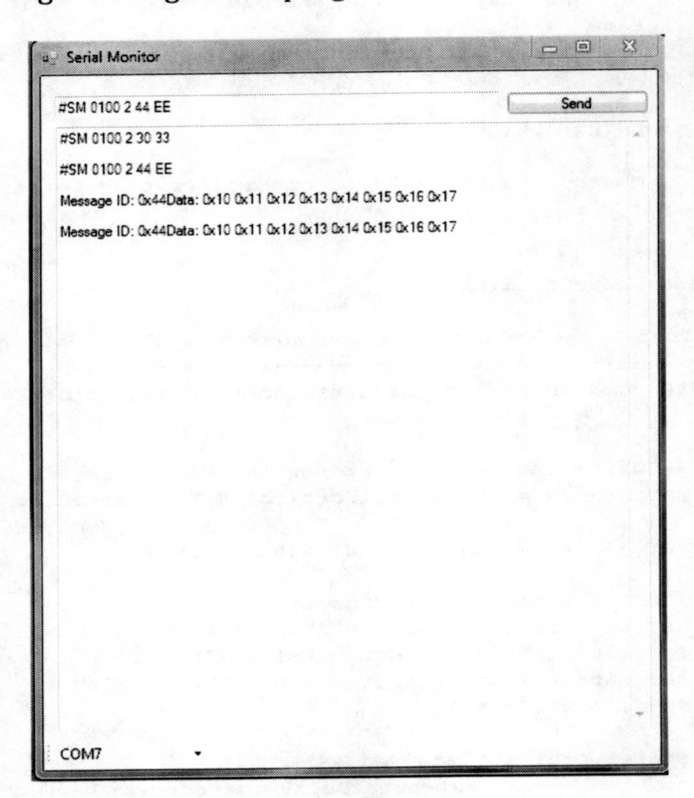

In this case, we sent two CAN Messages with the same ID (0100) but different data. Next, we received to messages through my standard test configuration (i.e. USB-to-CAN gateway with Windows monitoring tool as the second CAN node).

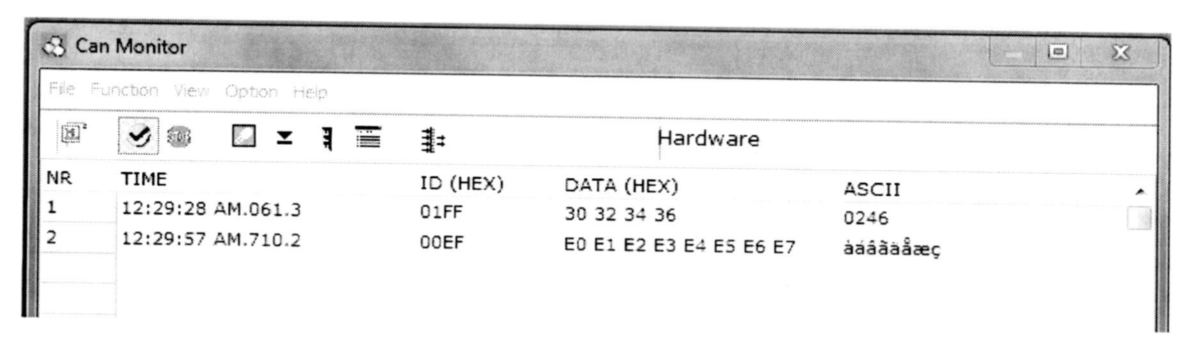

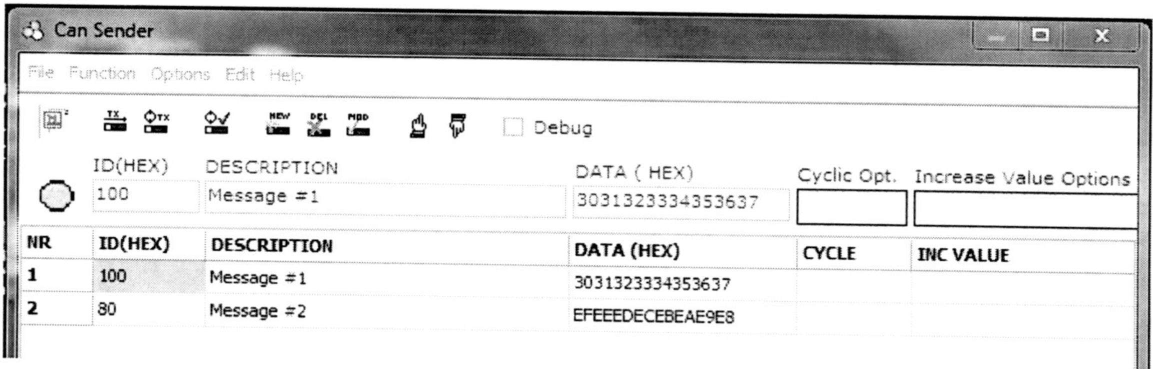

The previous two screen shots serve as evidence that the messages sent to/from the serial monitor via the Arduino CAN Shield were received/transmitted as expected.

5 Conclusion

At this point, after having accomplished all the necessary steps, it is easily possible to develop a professional, full-fledged CAN Network Monitoring, Diagnostics, and Simulation Software.

The groundwork has been laid for all necessary hardware and software components:

1. A USB-to-CAN Gateway to provide CAN connectivity to the host system
2. A communication protocol between the USB-to-CAN and the host system
3. A graphical user interface (GUI) for the presentation of the CAN network

For more technical information and articles and to contact me, see my website at http://copperhilltech.com.

Appendix – Recommended Literature

There is more than plenty and valuable literature available on the Arduino, but, being an experienced programmer, the one and only work I read was:

Programming Arduino
Getting Started with Sketches
By Simon Monk
ISBN 978-0-07-178422-1

Also recommended for providing more background information:

A Comprehensible Guide to Controller Area Network
By Wilfried Voss
ISBN 978-0976511601

A Comprehensible Guide to J1939
By Wilfried Voss
ISBN 978-0976511632

All works are available through Amazon.com and all their international online stores, Barnes & Noble, Abebooks.com and all their international online stores, and any other good book store.

CPSIA information can be obtained at www.ICGtesting.com
Printed in the USA
LVOW09s2335091215

466180LV00018B/537/P